SAE EDGE™
RESEARCH REPORT

Unsettled Topics in Automated Vehicle Data Sharing for Verification and Validation Purposes

Yaser Khalighi, Ph.D.
Caliber Data Labs, Inc.

Mohsen Khalkhali
Caliber Data Labs, Inc.

EDGE DEVELOPMENT TEAM

Michal Antkiewicz, Ph.D., *University of Waterloo*
Sven Beiker, Ph.D., *Silicon Valley Mobility*
Nicola Croce, *Deepen AI*
Martin Holder, *Institute of Automotive Engineering (FZD) at Technical University of Darmstadt*
Danny Kim, *VSI Labs*
Mohammad Musa, *Deepen AI*
Dr. Rahul Razdan, *Florida Polytechnic University*
Philipp Rosenberger, *Institute of Automotive Engineering (FZD) at Technical University of Darmstadt*
Evangelos Simoudis, Ph.D., *Synapse Partners*
Dr. Joachim G. Taiber, *International Alliance for Mobility Testing and Standardization*
Adrian Zlocki, Ph.D., *fka GmbH*

SAE INTERNATIONAL®

Warrendale, Pennsylvania, USA

About the Publisher

SAE International® is a global association of more than 128,000 engineers and related technical experts in the aerospace, automotive and commercial-vehicle industries. Our core competencies are life-long learning and voluntary consensus standards development. Visit sae.org

SAE EDGE™ Research Report Disclaimer

SAE EDGE™ Research Reports focus on topics that are dynamic, in which knowledge is incomplete, and which have yet to be standardized. They represent the collective wisdom of a group of experts and serve as a practical guide to the reader in understanding an unsettled subject matter. They are not meant to provide a recommended practice or protocol. The experts have assembled as a community of practitioners to contribute and collectivize their thoughts and points of view; these are not the positions of the institutions or businesses with which they are affiliated, nor is one contributor's perspective advanced over other contributors. SAE EDGE™ Research Reports are the property of SAE International and SAE alone is responsible for their content.

About This Publication

SAE EDGE™ Research Reports provide state-of-the-art and state-of-industry examinations of the most significant topics in mobility engineering. SAE EDGE™ contributors are experts from research, academia, and industry who have come together to explore and define the most critical advancements, challenges, and future direction in areas such as vehicle automation, unmanned aircraft, cybersecurity, advanced propulsion, advanced manufacturing, Internet of Things, and connectivity..

Related Resources

SAE MOBILUS® Automated & Connected Knowledge Hub
https://saemobilus.sae.org/automated-connected/

SAE Team

Frank Menchaca, Chief Growth Officer
Michael Thompson, Director, Standards, Information and Research Publications
Monica Nogueira, Acquisitions Director
Beth Ellen Dibeler, Product Manager
William Kucinski, Managing Technical Editor

Printed in USA

EPR2020007
ISSN 2640-3536
e-ISSN 2640-3544
ISBN 978-1-4686-0167-1

To purchase bulk quantities, please contact: SAE Customer Service

E-mail: CustomerService@sae.org
Phone: 877-606-7323 (*inside USA and Canada*)
+1-724-776-4970 (*outside USA*)
Fax: +1-724-776-0790

https://www.sae.org/publications/edge-research-reports

About the Editors

Yaser Khalighi, Ph.D. is a seasoned technologist and an entrepreneur. He earned a doctorate and master's degree from Stanford University with a specialization in high-performance distributed computing and mathematical modeling. His expertise includes artificial intelligence (AI), machine learning, and platform engineering.

Based in Silicon Valley, he has a stellar track record in building production-level computational solutions utilizing machine learning, simulation, and analytics to resolve a variety of complex, large-scale problems. He has been instrumental in recruiting and leading teams of engineers and scientists for reputable technology-based organizations.

Mohsen Khalkhali is a seasoned strategist, investor, and an entrepreneur. He has built and run global strategy and operations for a mobile telecoms company and started the world's first business-to-business transaction platform for telecom suppliers. More recently, he founded and ran a global venture capital firm investing in seed-stage startups with an active portfolio of 15 AI-rich companies across digital health, enterprise software-as-a-service, and industrial internet of things.

As a conscious leadership practitioner, Mohsen's expertise has been transformative in identifying, funding, and growing entrepreneurial teams that focus on exponential growth and long-term value.

contents

Unsettled Topics in Automated Vehicle Data Sharing for Verification and Validation Purposes

Abstract

The race to autonomy has been synonymous with the race to data collection. Automated vehicles (AVs) generate terabytes of data per day. Perception engineers use these large datasets to analyze and model the automated driving systems (ADS) that will eventually be integrated into vehicles that will drive themselves. However, the current industry practices of collecting data by driving on public roads to understand real-world scenarios is not practical and will be unlikely to lead to safe deployment of this technology anytime soon. Estimates show that it could take 400 years for a fleet of 100 AVs to drive enough miles to prove that they are as safe as humans.

We, therefore, discuss an unsettled topic of sharing data for verification and validation purposes where - instead of each testing project and organization doing their own tests in isolation and potentially duplicating work - a shared-data culture, business, and technology be developed. This could allow for rapid generation, testing, and sharing of the billions of possible scenarios that are needed to prove practicality and safety of an ADS - resulting in lower research and development costs to the industry. We explore how this could lead to better regulation, insurance, public acceptance - and finally, shorter technology development cycles. Finding a business case and changing to an open data culture are not going to be easy tasks, but we believe data sharing is the only way forward for the whole industry to move to the next phase of deployment after nearly a decade of intense research.

NOTE: SAE EDGE™ Research Reports are intended to identify and illuminate key issues in emerging, but still unsettled, technologies of interest to the mobility industry. The goal of SAE EDGE™ Research Reports is to stimulate discussion and work in the hope of promoting and speeding resolution of identified issues. SAE EDGE™ Research Reports are not intended to resolve the issues they identify or close any topic to further scrutiny.

YASER KHALIGHI, Ph.D.
Caliber Data Labs, Inc.

MOHSEN KHALKHALI
Caliber Data Labs, Inc.

Edge Development Team

Michał Antkiewicz, Ph.D., *University of Waterloo*
Sven Beiker, Ph.D., *Silicon Valley Mobility*
Nicola Croce, *Deepen AI*
Martin Holder, *Institute of Automotive Engineering (FZD) at Technical University of Darmstadt*
Danny Kim, *VSI Labs*
Mohammad Musa, *Deepen AI*
Dr. Rahul Razdan, *Florida Polytechnic University*
Philipp Rosenberger, *Institute of Automotive Engineering (FZD) at Technical University of Darmstadt*
Evangelos Simoudis, Ph.D., *Synapse Partners*
Dr. Joachim G. Taiber, *International Alliance for Mobility Testing and Standardization*
Adrian Zlocki, Ph.D., *fka GmbH*

ISSN 2640-3536

Introduction

By the end of 2020, there will be over 50 billion connected devices globally collecting over 2.3 zettabytes of data each year. Shared and open data gathered from these devices can underpin mobility solutions that support integrated, efficient, and sustainable transport systems. Automated vehicles (AVs) are estimated to generate 10 to 50 terabytes of sensor data each day alone. Considering that these vehicles are still in the test phase, the future possibilities and benefits of data sharing for autonomous-connected-electric-shared (ACES) vehicles are too large to ignore. These could be safer roads, enhanced mobility services, optimized transport networks, smart-city applications, improved security, and many more. We have focused our efforts in this report on sensor data used for perception aimed at verification and validation of automated driving system (ADS).

Public Road Testing and Validation

The current testing practices revolve around validating an ADS by running it through as many scenarios as possible. Scenarios involve a sequence of behavioral competencies in a specified driving environment that is often called an operational design domain (ODD). The composition of an ODD challenges the capabilities of the ADS. These scenarios are developed, tested, and validated in a closed process to provide feedback on system failures. It usually starts with description of common scenarios followed by multidimensional iterations of those scenarios (e.g., involving weather, time of day, location, speed, traffic patterns, etc.) in a virtual simulation environment. This is then followed by testing on closed tracks, and finally by testing on public roads with other road users. The nature of these tests is risky since the unknown unknowns expose the public to undue risk in the public-road testing phase. Additionally, AV developers may not want to push their systems to extremes in case of accidents.

According to a widely cited statistic by the RAND Corporation, at an 80 percent confidence level, AV developers need to drive 8.8 billion miles to validate that their ADS is capable of meeting the same level of human driver error of 1.09 fatalities per 100 million miles driven (Figure 1).

With a fleet of 100 AVs being test-driven 24 hours a day, 365 days a year, at an average speed of 25 miles per hour, this would take 400 years. This is clearly not reasonable, and no single developer can practically achieve the desired result. This is why we believe data sharing and collaboration in training, testing, and validation of these systems could be the answer.

Data Generation in Vehicles

The focus of this report is on the ADS that leverages perception sensor data to help an AV understand its environment and take the necessary driving action. However, vehicle data is nothing new. Connected vehicles are already generating large amounts of data, which we documented in Table 1 based on data sensitivity. This could be an indicator for the willingness of developers to keep the data open or closed addressing competition, user privacy, and security.

As we can see from Table 1, sensor data is classified as brand-differentiated data and hence the decision to make it open or closed may not be a trivial one.

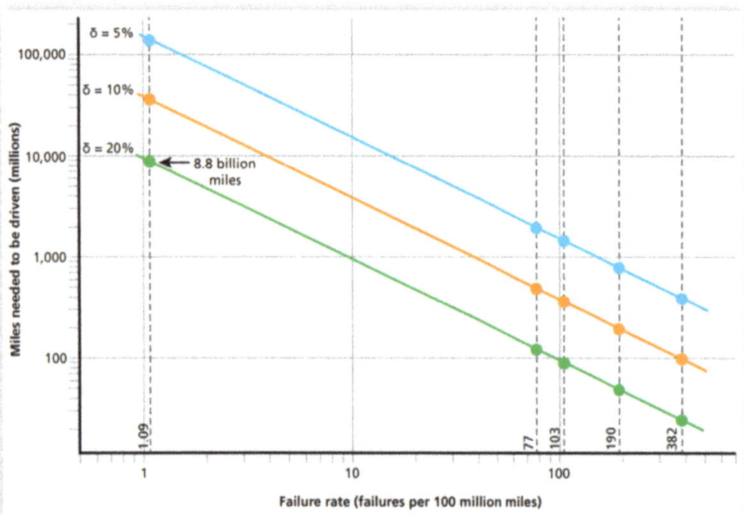

Reprinted with permission from Ref. [14]

FIGURE 1. **Miles needed to demonstrate failure rates to a particular degree of precision. These results use a 95-percent confidence interval. The three colored lines show results for different levels of precision δ, defined as the size of the confidence interval as a percentage of the failure rate estimate. The five dashed vertical reference lines indicate the failure rates of human drivers in terms of fatalities (1.09), reported injuries (77), estimated total injuries (103), reported crashes (190), and estimated total crashes (382). (Driving to Safety, RAND Corporation)**

TABLE 1. **Vehicle data classified by sensitivity.**

Nonbrand-differentiated data	Brand-differentiated data	Personal data
Data that is not differentiated by vehicle manufacturers; not considered "intellectual property sensitive"	Data that is differentiated by vehicle manufacturers; considered "intellectual property sensitive"	Data that supports services requiring user or vehicle identification
No data protection relevance as long as it is not tied to the vehicle identification number or any personal identifier	Strictly of a technical and/or operating nature	Data handling must meet strict data and privacy protection requirements
	No data protection relevance insofar as it is not tied to the vehicle identification number or any personal identifier	
	Examples	
Activation of hazard warning light	Lane marking perception	Vehicle location
Position of active emergency vehicles	Proprietary sensor data	Movement profile
Road conditions	Engine operating map	Average speed
Roadblocks and traffic flow data	Gearbox operating map	Acceleration
Ambient temperature	Actuator data	Fuel and consumption levels
Average speed	Engine injection behavior	Navigation destinations
On-street parking	Fuel pump performance	The user's address book
	Automatic transmission shifting behavior	Personalized access to third-party services
	Fault memory data	Infotainment settings
	Battery performance	Personalized in-car settings
	Stability control data	User's health and wellbeing data

Data Journey

A typical AV sees the world around it through multiple cameras, radar, and light detection and ranging (LiDAR) sensors. These sensors generate large amounts of data that need to be prepared and processed to develop models for an ADS. On a high level, an AV data journey consists of collection, management, labeling, and validation (Figure 2).

Collection (Sensors, Open Datasets, and Simulation)

The race to autonomy has become synonymous with the race to data collection. Our research led to the discovery of around 60 open datasets from both academia and industry. Table 2 displays details on each and where the datasets can be accessed.

Multiple companies are working on creating synthetic and enriched data for AVs to fill gaps for scenarios where measurement data is not available. Data-hungry perception systems need as much data as possible to create and test the edge cases required for training. Road testing remains the preferred route for data collection, testing, and validation due to the scenarios encountered in the real world, regardless of how often they occur. Furthermore, with regard to model validity, there are still no holistic simulation solutions available that can replace public road tests and data collection.

Storage (On-Premises, Private Cloud, and Public Cloud)

Once this large amount of data is generated and stored on the onboard hard drive, it needs to be moved for processing, model building, testing, and validation of those models. Moving this amount of data over the cloud can be costly - therefore, many of the AV companies have created custom ingestion stations where data is physically moved from the onboard storage to the offboard on-premises storage. Authors' calculations indicate that, for a fleet of 100 AVs, it takes 1 year to transfer a single day of data over a speed of 100 megabit per second and will cost $20 million annually to store on commercial cloud storage services.

FIGURE 2. **ADS data journey.**

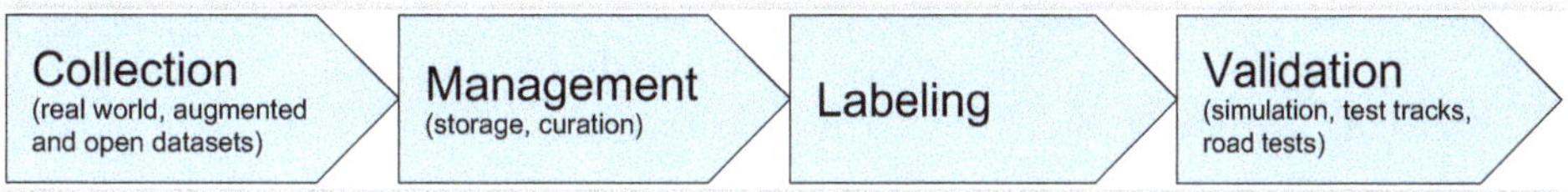

© SAE International.

TABLE 2. List and details on open datasets for AVs.

Name	Published	Type of data
The following databases are publicly accessible and free to view unless stated otherwise		
Apollo Open Data Platform	2018	Video, LiDAR, vehicle data, and bounding boxes
ApolloScape	2018	Scene parsing, car instance, lane segmentation, self-localization, trajectory, detection/tracking, stereo video data, and semantic label and lane marking
Argoverse by Argo AI	2018	Two public datasets with highly detailed maps, and three-dimensional (3D) tracking annotations for 113 scenes and 327,793 interesting vehicle trajectories for motion forecasting
Automotive multisensor dataset (AMUSE)	2013	Omnidirectional visual data for full surround sensing; include winter conditions with snow
Belgium Traffic Sign-ETH Zurich	2011	Images, bounding boxes
Bosch Small Traffic Lights Dataset	2017	Videos, bounding boxes
Boxy Vehicles Dataset-Bosch	2019	200,000 images, 1,990,000 annotated vehicles
Brain4Cars	2016	Video, Global Positioning System (GPS) data, vehicle data and codes, and behavioral labels
Caltech Pedestrian Detection Benchmark	2009	The largest pedestrian dataset on public roads, pedestrian annotation, the first dataset with temporal correspondence between bounding boxes and occlusion labels
CamVid-University of Cambridge	2009	Motion-based segmentation and recognition dataset: video, images, and codes, semantic labels
Carnegie Mellon University	2011	Various weather and light conditions
Cheddar Gorge	2011	Diversified sensor setup with stereo, monocular, and infrared cameras; Velodyne 64 LiDAR, GPS, inertial measurement unit, etc.
CityScapes	2016	Semantic understanding of 30 different classes in urban street scenes
CMU-Highway Workzones Dataset	2015	Videos, bounding boxes
Comma AI	2016	Video, GPS, vehicle data and codes; seven hours of largely highway driving
CCSAD	2014	Stereo video captured in developing countries
CULane Dataset	2018	Image and codes, lane marking
Daimler AG-LostandFound Dataset	2016	Image and vehicle data, semantic labels
Daimler Pedestrian Benchmarks Dataset	2006	Video and image, bounding boxes, and semantic labels
DAVIS Driving Dataset	2017	Video, GPS, vehicle data, and codes
DBNet-Large-scale Driving Behavior Dataset	2018	Video, LiDAR, and vehicle data
DIPLECS Autonomous Driving Datasets	2015	Video and vehicle data, behavioral labels
DReyeVE	2016	The first dataset for researching driver attention: human attention, eye fixation, and visual saliency
EISATS	2010	Includes multiple datasets with stereo vision sequences for comparative performance evaluation of stereo vision, optic flow, motion analysis, etc.
Elektra	2016	Various types of images with annotated pedestrians; including far infrared images
ETH Pedestrian Dataset	2009	Video, bounding boxes
Ford-Campus Vision and Lidar Dataset	2009	Video, LiDAR, GPS and codes, bounding boxes
German Traffic Sign Benchmarks Dataset	2012	Images, bounding boxes
Ground Truth Stixel Dataset-Daimler AG	2013	Video and vehicle data, Stixel
HCI Challenging Data for Stereo and Optical Flow	2012	Videos
HD1K Benchmark Suite	2018	Associated with an event called the Robust Vision Challenge; it provides challenging data for stereo and optical flow (e.g., rain flares and flying snow)
IEEE paper (This database is available for purchase.)	2017	Overview of 27 existing publicly available datasets
Joint Attention for Autonomous Driving (JAAD) Dataset	2016	Focus on joint attention between pedestrians and drivers for automated driving; provides both textual and behavioral annotations for pedestrians and vehicles

TABLE 2. **(Continued) List and details on open datasets for AVs.**

Name	Published	Type of data
Karlsruhe Dataset: Labeled Objects	2011	Images with object bounding boxes for cars and pedestrians; includes even object orientation
Karlsruhe Dataset: Stereo Video Sequences + rough GPS Poses	2009	High-quality stereo sequences in Karlsruhe
KAIST Multi-Spectral-Visual Perception for Autonomous Driving	2015	Video, LiDAR, GPS and thermal, bounding boxes
KAIST Urban Dataset	2018	Video, LiDAR, GPS and codes
KITTI Vision Benchmark Suite	2011	Stereo, optical flow, visual odometry, and 3D object detection and 3D tracking
levelXdata	2020	Unmanned aerial vehicle video data from drones
LISA (Laboratory for Intelligent & Safe Automobiles)-Traffic Sign Dataset	2012	Image and codes, bounding boxes
Lyft	2019	Large-scale dataset featuring raw sensor camera and LiDAR
The Málaga Stereo and Laser Urban Dataset	2013	Well-documented, full tool support with message boards embedded on its web page for convenient communication
Mapillary Vistas Dataset (This database is free for noncommercial use, but a license must be acquired for commercial use.)	2017	Images, semantic labels
Max Planck Institute for Informatics-Multi-Cue Onboard Pedestrian Detection	2009	Many different datasets: people detection, pose estimation and tracking, visual privacy, vision and language, human activity recognition, object recognition, and scene understanding
Motorway Dataset-Toyota Motor Europe	2011	28 clips for a total of approximately 27 minutes (30,000+ frames) with vehicle annotation
nuScenes-Aptiv, Scale & nuTonomy (This database is free for noncommercial use, but a license must be acquired for commercial use.)	2018	Large-scale automated vehicle dataset: scenes, images, LiDAR, radar, camera, objects by Aptiv
NVIDIA Self Driving Car Training Set	2017	1,000 images, mostly highway plus some urban driving
Oxford University-RobotCar Dataset (This database is free for noncommercial use, but a license must be acquired for commercial use.)	2015	Video, LiDAR, GPS, vehicle data and codes
PandaSet-Scale and HESAI (Soon to be released)	2019	Images and LiDAR, Bounding Boxes and Semantic Labels
Scale Driveability Dataset Index	Continually updated	The website hosts an up-to-date index of publicly available datasets for automated driving research
Semantic KITTI-University of Bonn	2019	Video, image, LiDAR, GPS and codes, bounding boxes, semantic label, and lane marking
Stanford Track Collection	2010	Velodyne 64-point cloud with object labels and GPS/IMU data
TU Darmstadt Visual Inference Group	2005	Various datasets
UC Berkeley DeepDrive (World's largest dataset)	2017	100,000 high-definition video sequences of over 1,100-hour driving experience, two-dimensional (2D) bounding boxes annotated on 100,00 images, 10,000 diverse images with pixel and instance level annotations
Udacity Open Source Self Driving Car	2016	Multiple datasets arranged in rosbag file format
University of Alcala Dataset	2016	500 minutes of naturalistic driving with its associated raw data and additional semantic information: video, GPS, vehicle data and codes, and behavioral labels
University of Malaga-Malaga Stereo and Laser Urban Dataset	2014	Video, LiDAR, GPS, and codes
Unsupervised Labelled Lane Markers Dataset-Bosch	2019	Over 100,000 annotated images, annotations of over 100 meters, LiDAR, and lane marking
Waymo	2019	3,000 driving scenes, 16.7 hours of video, 600,000 frames, 25 million 3D bounding boxes, 22 million 2D bounding boxes, five LiDARs, five cameras, and radars
York University-Joint Attention for Autonomous Driving (JAAD) Dataset	2016	Video, bounding boxes and behavioral labels

Curation (Search and Annotation)

Only when this data is summarized and indexed can it be viewed by perception engineers for scenario search and discovery. Most of the collected data is usually mundane highway driving where no new information is uncovered (Figure 3).

Many current machine learning approaches in ADS development are relying on supervised learning where annotation and labeling by humans is needed. This is a labor-intensive stage where human eyes need to go through hundreds of thousands of two-dimensional (2D) and three-dimensional (3D) images to classify objects.

Management (Share, Build, and Improve AI)

Once these objects are classified, they can be used by machine learning engineers and fed into neural networks (NNs), along with map/localization and trajectory planning data, to model driving decisions that a human normally makes in similar situations. This is highly valuable data that has undergone multiple stages of preparation and is seen as core intellectual property by AV companies. Sharing only happens internally, as anyone with access to the same data can train a similar model. However, with the addition of training data, models can be improved. So, even if data is shared at some point, there's a constant requirement for new data until models are validated to an acceptable level.

Validation (Test Tracks, Road Tests, Simulation)

Although there are some pockets of best practices developed, there is still no consensus on the accepted validation level. Every organization is going by a "best effort" principle and tries to self-certify the validity and safety of their own ADS.

Based on the National Highway Traffic Safety Agency (NHTSA) framework, two types of testing and validation are usual for ADS developers:

1. Challenges associated with AV technology: Focus on some of the characteristics of the technology and the underlying implementations of the integrated hardware and software systems. Examples of this are probabilistic and nondeterministic algorithms, machine learning algorithms, digital mapping needs, regression testing, and so on
2. Challenges associated with test execution: Highlight the expansiveness of the conditions that vehicles may encounter and handle with minimal, if any, input or guidance from a human. Examples of this are testing completeness, testing execution controllability, testing scalability, unknown or unclear constraints/operating conditions, degraded testing, infrastructure considerations, laws and regulations, and assumptions

These tests are done through three techniques:

1. Modeling and simulation (subsets: software-in-the-loop, hardware-in-the-loop, and vehicle-in-the-loop)
2. Closed-track testing
3. Open-road testing

These techniques offer a multifaceted testing architecture with varying degrees of test **control**, and varying degrees of **fidelity** in the test environment. There's still no best practice on how to go from data collection to open-road testing and validation. We have provided a high-level comparison of each testing techniques in Table 3:

Recommendations

1. Develop clear rules around privacy, security, and data ownership

FIGURE 3. **Not all data is of the same quality when building ADS models.**

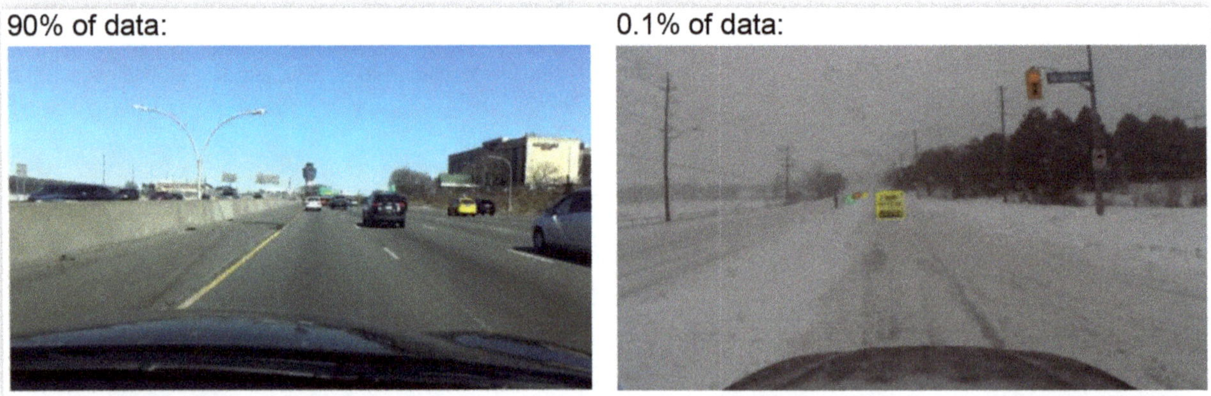

Reprinted with permission from Berkeley DeepDrive/University of California: Berkeley; Reprinted with permission from Canadian Adverse Driving Conditions Dataset

TABLE 3. High-level comparison of common testing and validation techniques.

High = 3, Low = 1	Simulation	Track testing	Road testing
Controllability	3	2	1
Predictability	3	2	1
Repeatability	3	3	1
Scalability	3	2	1
Efficiency	3	1	1
Fidelity	1	2	3
Transferability	3	3	1
Standardize output	1	2	3
Data sharing capability	2	3	1
Cost-effectiveness	2	2	1
Variability	3	2	1
Safety	3	2	1

2. Develop industry-wide standards for the sensor data, sensor data format, labels, and scenario description language
3. Develop collaborative collection efforts based on scenarios and gaps in the training data

Data Sharing, Safety, and Incentives

Shared data is data that a specific person or group of people, who are not the data subject, owner, or holder can access. There are three main types of shared data:

1. **Public access** is data that is available to anyone under certain terms and conditions.
2. **Group-based access** is data that is available to specific groups of people or organizations who meet certain criteria.
3. **Named access** is data that is shared only with named people or organizations.

AVs will radically transform the way data is shared and protected as these software- and hardware-enabled vehicles will generate large amounts of it. The challenge will be in determining who owns the data, how to use it, and how to protect it. Ensuring data protection and cybersecurity is one of the key priorities outlined at the G7 declaration on automated and connected driving. Global Positioning System (GPS) devices already collect data, and with the growth of AVs, we could see the expansion of related data, including passenger identities, personal profiles, origins and destinations, travel dynamics, media use, and so on. For example, a code of practice in the United Kingdom for testing driverless cars specifies the recommended actions and practices for AV firms related to data collection, data protection, cybersecurity, and software management. Similarly, the United States' NHTSA includes data sharing as part of their 12 elements of safety that AV developers could voluntarily adhere to. **Unsettled topics such as data privacy will benefit greatly from more work and discussion among government, industry, and stakeholders.** Along with managing public concerns related to data privacy protection, regulators need to determine who will have access to data, and on what conditions data will be shared.

It's probably fair to say that automakers have been competing on safety for 100 years and are only now learning how to collaborate on safety. Most of the collaboration and effort is around building test and validation specifications to address things like "why it's important to vet the behavior of an ML-driven ADS and make sure it is acting safely," even if there is no fault in its hardware and no bugs in its software. However, in the current industry practices, sharing is still a long way away from showing meaningful results.

Government-funded projects in Europe, such as ConVEx, have made data sharing and brokerage their mission with results that are yet to be published.

There are three core reasons why mobility and transportation organizations lack incentives to share data:

1. The fear that shared data could lead to breaches in privacy, security, and safety
2. The belief that the costs of sharing data outweigh the benefits
3. The focus by organizations on their own mode of transport, limiting opportunities, and awareness to make data available beyond their segment

Several industry stakeholders refer to the high costs involved in securing access to data or making it available to others. The typical length of time required to negotiate access to data covering issues including ownership, licensing, and revenue-sharing issues can be long and distracting for a young and growing industry. This cost is particularly acute for small- and medium-sized businesses, which cannot afford legal advice to navigate through complex agreements. Given that value is often derived from combining multiple datasets, the barriers to acquisition for each dataset, and abiding to the terms of these agreements makes it hard to create and test innovative products. From a data engineering point of view, the high cost of making data available in a shareable format can be another reason data sharing can be a high-cost practice with low (or unknown) benefits.

What Is the Business Case?

While there is a balance to be achieved between investing in sharing data - and a potential or perceived commercial loss from sharing - not exploring these trade-offs will mean data remains locked inside individual organizations and not available for use in new mobility solutions.

The focus of this SAE EDGE™ Research Report is on public access data or group access, where an independent mechanism exists to incentivize organizations to share data that would eliminate the cost/benefit concerns.

Cities and transportation authorities often have a tendency to share and open their data so that new mobility services can be built on top of aggregated movement data. Safety-related data is usually shared between law enforcement, licensing, and operators of mobility services so that safety practices are shared across the industry as quickly as possible.

However, development data in the form of raw sensor data is seen as a proprietary asset that is only shared in public projects. These are funded by governments or academic institutions, or consortia between private and public companies where sharing is frequently a matter of long discussions and tension between consortium partners. **The unsettled topic of incentivizing private organizations to share their data, and - more specifically, scenarios - to verify and validate an ADS has been so far unaddressed by the industry.** It is usually seen as a regulatory burden where the cost of not sharing directly translates to fines or reputational damage in terms of unsafe technology.

Two possible business models for sharing development data could be:

1. **Data exchange:** In this membership model, members get rewarded for sharing new edge cases and the central platform assists in legal processes and pricing these two-party or multiparty contracts.
2. **Validation as a service:** This includes continuous validation of models against open datasets, simulation data, test-track data, and road-test data where recommendations on gaps in training data are monetized.

We believe that in the short term, data sharing can reduce research and development costs, and in the medium to long term, it can lead to an accelerated deployment of the technology.

What Are the Technical Implications of Data Sharing?

While it is quite common to utilize open-source data to build proof-of-concept grade perception systems, there are several technical barriers in sharing driving data between teams that build production-grade solutions. These difficulties largely arise from the complexity of perception models and their dependence on the vehicle configuration. We can categorize these technical difficulties into the following.

Sensor Configuration Perception models are sensitive to the specifics of the sensors of which the training data is collected from. For example, in the case of an object detection deep-learning model, if the training data originated from a camera with significantly different calibration than the camera installed on the production vehicle, the object detection module may perform poorly. In this case, the safest option for a perception engineer is to obtain the training data from a camera with the same intrinsic (e.g., focal length, principal point) and extrinsic settings (e.g., camera location, angle). This poses a problem in data sharing, because various AV development teams have different sensor configurations and calibrations. In addition, some perception tasks are more sensitive to calibrations than others. For example, a deep-learning algorithm that estimates the distance of objects from the vehicle is far more sensitive to the focal distance of the camera than an object-detection model that only identifies the type of traffic objects. That means without the knowledge of sensor configurations, particularly the intrinsic and extrinsic characteristics, the dataset generated from a sensor might be of little to no value to an external team for model-development purposes.

Having said that, it is a common practice for vision engineers to enrich their training datasets with open-source data and/or use transfer-learning techniques. In these cases, learned knowledge is transferred from one system to another by assuming some generic features, which are shared across multiple domains. As a result, depending on the model training techniques, there is *some* value to be gained from obtaining external data, even if there is a sensor mismatch.

Labeling Specification AV developers commonly use supervised learning to train the perception models, which means the raw data obtained from sensors is not useful by itself. The raw data needs to be labeled by human labelers. The perception/algorithm engineers that architect the perception model also design and provide specific instructions for labeling teams to create labeled datasets custom-tailored to the architected model. The specificity of these labeling requirements to the model is another source of difficulty for sharing labeled data. For example, take an object detection module that detects bikes and pedestrians. In a scenario where a pedestrian is walking with a bike, one team may decide to combine the pedestrian and the bike and label them as one object, while another team may distinguish them as two separate objects. Combining these mismatched labeled datasets for training could result in poor detection of bike-walkers by the trained model.

Common Standard for Data Evaluation and Description AV developers need to discover the gaps in their development datasets to find what data they need to acquire. This requires a well-defined, industry-wide comprehensive set of driving scenarios and road conditions such that the model is tested extensively, its weak points are discovered, and the performance is improved by importing external data. Such industry-wide standards for driving scenarios still do not exist. In addition, the industry should adopt a common data description and exchange language. On top of that, the Association for Standardization of Automation and Measuring Systems (ASAM) has started the development of a set of standards - called OpenX - aimed at tackling these issues. Foretellix, a startup focused on safety validation of ADS, has made considerable contribution in this effort by introducing Measurable Scenario Description Language, or "M-SDL," which is a high-level language for describing driving scenarios. However, the industry is still far from widely adopting such common frameworks.

Recommendations

1. Incentivize private companies to share data and protect them against unwanted consequences of data sharing.
2. Standardize data exchange practices and open application programmable interfaces for seamless data exchange.
3. Develop common validation metrics so that an ADS can be validated by third parties as well as AV developers.

Safety, Regulation, Liability, Public Acceptance, and Culture

Data is essential to realize the vision of a future transport system that meets the expectations and exploits the capabilities of the internet age. Establishing a sharing culture is by far the most effective way governments, regulators, and society can help new technologies develop. To help people and goods move easily, cheaply, and efficiently across every form of transportation, we need data to flow freely too. That data can help everyone make better decisions, including passengers, freight companies, transport operators, and policymakers. Over the last few years, we have learned that when we make data as open as possible, it maximizes the creation of value and unleashes innovation. Data is unlike many other assets. The benefits from its use and reuse grow exponentially the more it is shared and open, combined, and linked with other data. The benefits of making this change to sharing and openness are widely recognized, but we don't always talk about the opportunity cost of doing nothing. For example (according to a March 2017 briefing paper "The case for government involvement to incentivise data sharing in the UK Intelligent Mobility sector"), the United Kingdom has estimated that not sharing and making more transport data open could cost the country £15 billion by 2025. This will mean poorer quality services, reduced transport connectivity, and a lost opportunity to use mobility innovations as a driver for economic growth and social change.

Higher levels of road-vehicle automation pose a regulatory challenge across the globe. In the United States, at the national level, NHTSA has been responsible for ensuring vehicle safety through the mandatory Federal Motor Vehicle Safety Standards and the voluntary New Car Assessment Program. Although NHTSA typically regulates the technology aspects of vehicle safety, state agencies-such as the state Departments of Motor Vehicles, are responsible for the regulations governing training, evaluation, and licensing of drivers and registration of vehicles. Automation that allows drivers to disengage from monitoring and control tasks introduces safety concerns related to the vehicle technology (typically regulated by NHTSA) and to the automated vehicle's driving behavior and compliance with vehicle codes (typically regulated by individual states). **Issues that need to be addressed in support of state regulations for manufacturers' testing of AVs on public roads and the general public's operation of ADS are greatly discussed but remain**

FIGURE 4. The virtuous cycle of data sharing and safer roads.

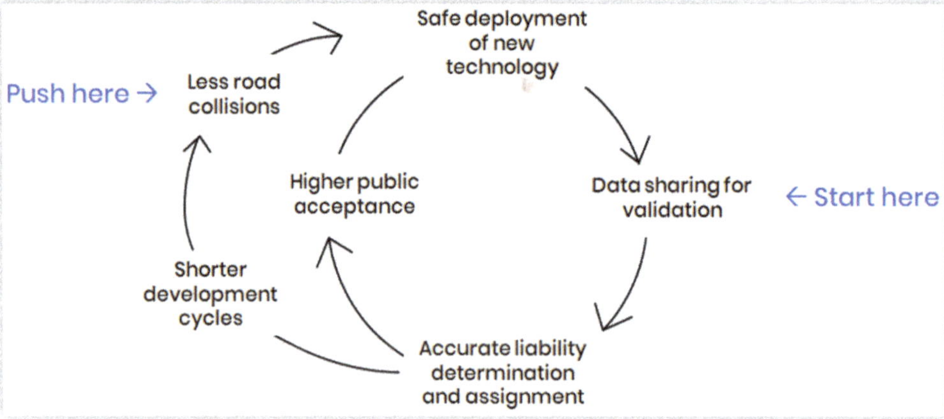

unsettled topics. The key challenges for these regulations are how to ensure public safety without discouraging technical innovations, and how to define meaningful requirements in the absence of existing technical standards for ADS.

We believe there exists a virtuous cycle of data sharing, regulation, public acceptance, shorter innovation cycles, and - most importantly - safer roads that can be used to guide discussions about the unsettled domain of data sharing (Figure 4).

How Does Data Help Regulators Verify the Safety of AVs?

Drawing from a real AV testing crash that resulted in a tragic human fatality, we refer to a recent report by the US National Transportation Safety Board (NTSB) on Uber's fatal crash in March 2018 where they made the following recommendations:

To NHTSA

1. Require entities who are testing or who intend to test a developmental ADS on public roads to submit a safety self-assessment report to your agency.
2. Establish a process for the ongoing evaluation of the safety self-assessment reports as required in Safety Recommendation and determine whether the plans include appropriate safeguards for testing a developmental automated driving system on public roads, including adequate monitoring of vehicle operator engagement, if applicable.

To the state of Arizona

3. Require developers to submit an application for testing ADS-equipped vehicles that, at a minimum, details a plan to manage the risk associated with crashes and operator inattentiveness and establishes countermeasures to prevent crashes or mitigate crash severity within the ADS testing parameters.
4. Establish a task group of experts to evaluate applications for testing vehicles equipped with automated driving systems, as described in Safety Recommendation 3, before granting a testing permit.

To the American Association of Motor Vehicle Administrators

5. Inform the states about the circumstances of the Tempe crash and encourage them to (1) require developers to submit an application for testing ADS-equipped vehicles that, at a minimum, details a plan to manage the risk associated with crashes and operator inattentiveness and establishes countermeasures to prevent crashes or mitigate crash severity within the ADS testing parameters, and (2) establish a task group of experts to evaluate the application before granting a testing permit.

To the Uber Technologies, Inc., Advanced Technologies Group

6. Complete the implementation of a safety management system for automated driving system testing that, at a minimum, includes safety policy, safety risk management, safety assurance, and safety promotion.

The chairman of NTSB admitted that only 20 percent of AV manufacturers have submitted voluntary safety self-assessment. In most of these safety assessments, data sharing and reporting are topics of discussion, but no real operational plan or practical way of sharing is discussed or outlined.

The United Kingdom, through its Centre for Connected and Autonomous Vehicles, also recommends that testing organizations should issue an abridged safety case for the public, but only leaves this as a voluntary practice.

Clearly, more needs to be done. Currently, AV manufacturers lack the urgency to share more data. Sharing testing data will increase visibility into safety practices. If the industry wishes to move from development to deployment successfully, data sharing and collaborative learning needs serious thought and action. Similar to the way that the aviation industry has built an enviable safety reputation by sharing safety best practices, the automated driving industry can follow in their footsteps and not reinvent the wheel.

How Does Data Help Insurers Verify the Safety of AVs?

Insurers say they currently do not have enough data to validate the auto industry's promises of safety benefits from ADS. They cite automakers' reluctance to provide detailed information on models sold with those features, their lack of consistent standards, drivers' unpredictable use of the systems, and higher repair costs. There are even major inconsistencies when it comes to terminology of ADS features. In a January 2019 report titled "Advanced Driver Assistance Technology Names," the American Automobile Association identified at least 20 unique names for adaptive cruise control alone!

A report by the Insurance Bureau of Canada (an industry association) recommends these elements of data to be shared for insurance purposes:

1. GPS-event time stamp
2. GPS-event location
3. Automated status-on or off
4. Automated mode-parking or driving
5. Automated transition time stamp
6. Record of driver intervention of steering or braking, throttle or indicator
7. Time since last driver interaction
8. Driver seat occupancy
9. Driver belt latch
10. Speed
11. Vehicle warnings or notifications to the vehicle's operator

Thatcham Research, the United Kingdom's insurance industry vehicle safety and research center, deems the first nine data elements necessary to support the single insurance policy prescribed in the country's Automated and Electric Vehicles Bill. The nine data elements should help confirm whether the vehicle was in manual or automated mode at the time of the collision.

Without standardization of the basic terminology, understanding the correct hand-off procedures, and determination of liability, it remains difficult to come up with mass market insurance policies for AVs, and insuring them remains a bespoke practice. Data sharing could allow insurers a better understanding of the validation process and even allow them to help influence product development so that liability determination can occur efficiently.

Open-Source Culture for Sharing

In the past decades, software companies relied heavily on trade secrets for creating competitive, high-quality products. Nonetheless, in the last few years, software companies have recognized the value of open-source software development and the opportunity for superior quality that can be created by the developer community. However, this has also led to the insight that in-house developments often bear no relation to the adoption of open-source solutions. The same trend could take shape in data sharing for AV development. The marriage of the modern software culture of transparency with the old-world culture of the original equipment manufacturer (OEM), now means that the two cultures must mesh to find a new balance. The same way that Microsoft has started to embrace the open-source software by discovering the value and quality in open-source development, we believe that OEMs will discover the value of data sharing and accept the collaboration culture in their AV development.

Projects Promoting a Culture of Data Sharing

In our research, we came across multiple projects that promote data sharing. Projects, such as ConVEx in the United Kingdom, focus on connected-vehicle data sharing. Organizations such as ASAM work on frameworks for sharing driving scenarios for simulation purposes. Global initiatives, such as Safety Pool, gather multiple players to share data for AV safety.

Safety Pool The World Economic Forum and Deepen AI, a Silicon Valley startup specializing in AI-powered software tooling for autonomous systems development, have launched a relevant initiative aimed at realizing an impactful data-sharing vision for the automotive industry. Safety Pool is a global brokerage allowing each individual organization developing SAE International Level 2+/3 Advanced Driver Assistance Systems features or Level 4+ ADSs to share driving scenarios, safety, and even raw sensor data under a common hood. An incentive scheme ensures that every participant contributing to the Safety Pool gets rewarded with access to valuable data under a framework that maximizes benefits, regardless of the participants' organization size, funding situation, or years of operations.

The rationale behind this initiative is based on the acknowledgment that the lack of data sharing is holding back the whole ADS industry while generating negative externalities. Lack of sharing makes each company less capable of dealing with peculiar situations generated by their unique data gaps, diminishing the ability for different ADSs to cope with each other in critical situations, and preventing the development of standards.

The actors and stakeholders suffering consequences are not only organizations developing the technology, but also regulators and policymakers who do not have access to common foundations to build benchmarks and data-driven guidelines. Most importantly, road users and consumers, who are unnecessarily exposed to higher risks, won't have access to automated mobility as fast as they would if data sharing was mainstream.

In summary, the Safety Pool vision is to drastically increase overall operational safety through virtuous collaboration, while helping the whole AV industry move forward, closer to deployment faster.

Recommendations

1. Embrace a culture of sharing, like that of open-source software.
2. Involve more stakeholders, such as regulators and insurance companies, in the validation process by sharing data that could help them understand and interact with the technology better.
3. Develop a clear separation between product-related proprietary data and shareable safety-related data.

Summary

For the development and validation of ADS technology, sharing data is not optional, it's a necessity. Otherwise, it will take decades (if not more) and cost billions of dollars to develop acceptable AV solutions that deliver the promised benefits of reduced fatalities, increased mobility access, and higher productivity. As we outlined above, there is no shortage of obstacles in the journey to data sharing such as culture, privacy, technical, and interoperability barriers. However, we remain optimistic that a paradigm shift from industrial-age thinking of trade secrecy to a knowledge-economy mindset will occur.

SAE EDGE™ Research Reports

SAE EDGE™ Research Reports, like the present report on "Unsettled Topics in Automated Vehicle Data Sharing for Verification and Validation Purposes," are intended to push further out into still unsettled areas of technology of interest to the mobility industry. SAE launches these reports before attempting to form a joint working group, let alone a cooperative research program or a standards committee.

SAE EDGE™ reports are intended to be quick, concise overviews of major unsettled areas where vital new technologies are emerging. By definition, an unsettled area is characterized more by confusion and controversy than established order. Early practitioners must confront an absence of agreement. Their challenge is often not to seize the high ground but to find common ground. These scouting reports from the frontiers of investigation are intended merely to begin the process of sorting through critical issues, contributing to a better understanding of key problems, and providing helpful suggestions about possible next steps and avenues of investigation.

SAE EDGE™ Research Reports, therefore, are fundamentally distinct from the more formal working groups approach and far removed from the more mature CRP and standard's development process.

Next Steps for Automated Vehicle Data Sharing for Verification and Validation Purposes

This publication should be considered only as a first step toward clarifying the issues around automated vehicle data sharing for verification and validation purposes. The intention behind this and other SAE EDGE™ Research Reports is to start a dialogue among interested parties on important industry-wide topics that require further attention. The expectation is that these explorations of unsettled areas of technology will lead to the formation of working groups and, ultimately, committees that can address and resolve the issues they raise, producing a framework for developing a common vocabulary of definitions, best practices, protocols, and standards needed to support continued progress toward safer and more innovative automotive products.

The experts' collaboration that gave rise to this publication demonstrated a great willingness on the part of the industry to define the terminology, procedures, and eventually the standards needed to enable automated vehicle data sharing technology to move ahead as quickly and efficiently as possible. SAE International has demonstrated its lead in this and closely related areas.

This SAE EDGE™ Research Report on "Unsettled Topics in Automated Vehicle Data Sharing for Verification and Validation Purposes" identifies the following key topics for further pursuit, both through continued informal discussions among industry practitioners and through more formal working groups:

- The need for leadership on safe testing approaches specifically public road testing
- The need for early collaboration on data ownership understanding and regulation
- The need to establish definitions on type of data collected from many vehicle systems
- The need to address cybersecurity and data storage and access early on

Recommendations

The overall recommendations of this SAE EDGE™ Research Report can be summarized as follows:

- Establish minimum data requirement standards to safely develop and validate an AV system. In particular
  - Expand the work of ASAM from simulation to test-track and road testing validation.
  - Enforce standards driven by international bodies such as the International Alliance for Mobility Testing and Standardization established by SAE International.
- Encourage collaboration through testbed data sharing as natural mobility innovation hubs.
- Work closely with the insurance industry to understand safety and liability implications from the earliest design and development stages.
- Produce public ODD data to be maintained by cities, municipalities, and local authorities.

Abbreviations/Definitions

2D - Two-Dimensional

3D - Three-Dimensional

ACES - Autonomous-Connected-Electric-Shared

ADS - Automated Driving Systems

AI - Artificial Intelligence

ASAM - Association for Standardization of Automation and Measuring Systems

AV - Automated Vehicle

GPS - Global Positioning System

LiDAR - Light Detection and Ranging

NHTSA - National Highway Traffic Safety Agency

NN - Neural Networks

NTSB - National Transportation Safety Board

OEM - Original Equipment Manufacturer

Radar - Radio Detection and Ranging

SAE J3016™ automated driving levels:

L0 - Level 0, or no driving automation

L1 - Level 1, or driver assistance

L2 - Level 2, or partial driving automation

L3 - Level 3, or conditional driving automation

L4 - Level 4, or high driving automation

L5 - Level 5, or full driving automation

Acknowledgements

Recognition should go first to all the participants, many of who also provided feedback on the draft version of this publication. Without their input and initiative, this SAE EDGE™ Research Report would not have been possible.

Sven Beiker, Ph.D., Silicon Valley Mobility
Dr. Rahul Razdan, Florida Polytechnic University
Evangelos Simoudis, Ph.D., Synapse Partners
Danny Kim, VSI Labs
Adrian Zlocki, Ph.D., fka GmbH
Martin Holder, Institute of Automotive Engineering (FZD) at Technical University of Darmstadt
Philipp Rosenberger, Institute of Automotive Engineering (FZD) at Technical University of Darmstadt
Dr. Joachim G. Taiber, International Alliance for Mobility Testing and Standardization
Michał Antkiewicz, Ph.D., University of Waterloo
Mohammad Musa, Deepen AI
Nicola Croce, Deepen AI

The authors of this document together with the SAE Team responsible for its creation join in expressing our deepest appreciation to all the individuals mentioned above.

Yaser Khalighi, Caliber Data Labs, Inc.
Mohsen Khalkhali, Caliber Data Labs, Inc.

References

1. Bellon, T., "Tesla's Use of Individual Driver Data for Insurance 'state-by-state proposition," Reuters, September 4, 2019, https://www.reuters.com/article/us-tesla-markel-insurance/teslas-use-of-individual-driver-data-for-insurance-state-by-state-proposition-idUSKCN1VQ0FY, accessed September 5, 2019.
2. Sanders, G. and Kaul, A. "Robotics Use Cases," Tractica, December 20, 2018, https://www.tractica.com/research/robotics-use-cases/, accessed September 5, 2019
3. Canadian Council of Motor Transport Administrators, "Automated Vehicles in Canada," November 20, 2016, http://www.ccmta.ca/images/publications/pdf/CCMTA_WhitePaper_Automated_Vehicles_in_Canada_2017.pdf, accessed September 2019.
4. Fogarty, K., "How Many Test Miles Make A Vehicle Safe?" Semiconductor Engineering, August 6, 2019. https://semiengineering.com/how-many-test-miles-make-a-vehicle-safe/, acessed September 5, 2019.
5. Nowakowski, C., Shladover, S., and Tan, H., "Heavy Vehicle Automation: Human Factors Lessons Learned," in *Elsevier BV. 6th International Conference on Applied Human Factors and Ergonomics (AHFE 2015) and the Affiliated Conferences, AHFE 2015*, Richmond, CA, 2015, https://doi.org/10.1016/j.promfg.2015.07.824.

6. Salay, R., Queiroz, R., and Czarnecki, K., "An Analysis of ISO 26262: Machine Learning and Safety in Automotive Software," SAE Technical Paper 2018-01-1075, 2018, https://doi.org/10.4271/2018-01-1075.
7. "The Case for Government Involvement to Incentivise Data Sharing in the UK Intelligent Mobility Sector," Catapult Transport Systems, March 17, 2017, https://s3-eu-west-1.amazonaws.com/media.ts.catapult/wp-content/uploads/2017/04/12092544/15460-TSC-Q1-Report-Document-Suite-single-pages.pdf, accessed September 6, 2019.
8. Cisco, "VNI Global Fixed and Mobile Internet Traffic Forecasts," March 20, 2019, https://www.cisco.com/c/en/us/solutions/service-provider/visual-networking-index-vni/index.html, accessed October 1, 2019.
9. Rossi, T. "Autonomous and ADAS Test Cars Produce over 11 TB of Data per Day," Tuxera, https://www.tuxera.com/blog/autonomous-and-adas-test-cars-produce-over-11-tb-of-data-per-day/, accessed October 1, 2019.
10. Insurance Bureau of Canada, "A Recommended Canadian Insurance Framework for Automated Vehicles," 2018, http://assets.ibc.ca/Documents/The-Future/Automated-Vehicle-Report.pdf, accessed November 18, 2019.
11. National Transportation Safety Board, "Collision between Vehicle Controlled by Developmental Automated Driving System and Pedestrian, Tempe, Arizona," March 18, 2018, https://www.ntsb.gov/news/events/Documents/2019-HWY18MH010-BMG-abstract.pdf, accessed November 19, 2019.
12. Rodrigo, C., "Senators Grill Safety Regulator over Self-Driving Cars," The Hill, November 20, 2019, https://thehill.com/policy/technology/471382-senators-grill-safety-regulator-over-self-driving-cars, accessed November 20, 2019.
13. UK Department for Transportation and the Centre for Connected & Autonomous Vehicles, "Code of Practice: Automated Vehicle Trialing," February 2019, https://assets.publishing.service.gov.uk/government/uploads/system/uploads/attachment_data/file/776511/code-of-practice-automated-vehicle-trialling.pdf, accessed November 20, 2019.
14. Kalra, N. and Paddock, S., "Driving to Safety," RAND Corporation, 2016, https://www.rand.org/content/dam/rand/pubs/research_reports/RR1400/RR1478/RAND_RR1478.pdf, accessed: January 10, 2020.
15. American Automobile Association, "Advanced Driver Assistance Technology Names," January 2019, https://www.aaa.com/AAA/common/AAR/files/ADAS-Technology-Names-Research-Report.pdf, accessed November 18, 2019.

Contact Information

EDGEresearch@sae.org

The authors of this document together with the SAE Team responsible for its creation join in expressing our deepest appreciation to all of the individuals who contributed.

www.ingramcontent.com/pod-product-compliance
Lightning Source LLC
LaVergne TN
LVHW070913210926
848482LV00017B/1064
9781468601671